→INTRODUCING

GAME THEORY

**IVAN PASTINE, TUVANA PASTINE
& TOM HUMBERSTONE**

KT-219-983

This edition published in
the UK and the USA in 2017
by Icon Books Ltd,
39-41 North Road, London N7 9DP
info@iconbooks.com
www.introducingbooks.com

Sold in the UK, Europe and Asia
by Faber & Faber Ltd,
Bloomsbury House,
74–77 Great Russell Street,
London WC1B 3DA or their agents

Distributed in the UK, Europe and
Asia by Grantham Book Services,
Trent Road, Grantham, NG31 7XQ

Distributed in South Africa
by Jonathan Ball, Office B4,
The District, 41 Sir Lowry Road,
Woodstock 7925

Distributed in Australia and
New Zealand by Allen & Unwin Pty
Ltd, PO Box 8500, 83 Alexander
Street, Crows Nest, NSW 2065

Distributed in the USA
by Publishers Group West
1700 Fourth Street
Berkeley, CA 94710

Distributed in Canada
by Publishers Group Canada,
76 Stafford Street, Unit 300,
Toronto, Ontario M6J 2S1

Distributed in India
by Penguin Books India, 7th Floor,
Infinity Tower – C, DLF Cyber City,
Gurgaon 122002, Haryana

ISBN: 978-178578-082-0

Text and illustrations copyright © 2017 Icon Books Ltd

Editor: Kiera Jamison

Printed in Great Britain by Clays Ltd, Elcograf S.p.A.

What is game theory?

Game theory is a set of tools used to help analyze situations where an individual's best course of action depends on what others do or are expected to do. Game theory allows us to understand how people act in situations where they are interconnected.

Connections between people arise in all sorts of situations. Sometimes through **cooperation** with others we can achieve more than we can on our own. Other times **conflict** arises where an individual benefits at the expense of others. And in many situations, there are benefits to cooperation but elements of conflict also exist.

We only win when we work together, but right now everybody is trying to be the star.

Because game theory can help analyze any environment where a person's best action depends on others' behaviour, it has proven useful in a wide variety of fields.

In *economics*, the decisions of firms are affected by their expectations of a competitor's choice of product, price and advertising.

In *political science*, a candidate's policy platform is influenced by the policy announcements of their rival.

In *biology*, animals must compete for scarce resources, but can be hurt if they are too aggressive with the wrong rival.

In *computer science*, networked computers compete for bandwidth.

In *sociology*, public displays of non-conformist attitudes are influenced by others' behaviour, which is shaped by social culture.

Game theory is useful whenever there is **strategic interaction**, whenever how well you do depends on the actions of others as well as your own choices. In these cases, people's actions are influenced by their expectations of others' actions.

Why is it called "game theory"?

Game theory is the study of strategic interaction. Strategic interaction is also the key element of most board games, which is where it gets its name. Your decision affects the other player's actions and vice versa. Much of the jargon of game theory is borrowed directly from games. The decision makers are called **players**. Players make a **move** when they make a decision.

Working with models

Real-world strategic interaction can be very complicated. In human interaction, for instance, it's not just our decisions, but also our expressions, our tone of voice and our body language that influence others. People bring different histories and points of view to their dealings with others. This infinite variety can create very complex situations that are difficult to analyze.

We can circumvent this complexity by creating simplistic structures, called **models**. Models are simple enough to analyze but still capture some important feature of the real-world problem. A cleverly chosen simple model can help us learn something useful about the complex real-world problem.

The game of chess is useful for understanding the complexity that variation brings to playing (and to predicting) games and outcomes. There are well-defined rules in chess. There are a limited number of options in each move. Yet the complexity of the game is daunting even though it is much simpler than even the most basic human interaction.

"It's a draw."

One feature of complex board games like chess is that the more skilled the players are, the more frequently the game ends with a draw. How can we explain this observation?

Since chess itself is too complex to fully analyze, let's use a simple model that captures some of the important features of the chess game: noughts & crosses (tic-tac-toe). Both chess and noughts & crosses have well-defined boards and victory conditions. Players take turns making choices from a limited selection of possible moves.

There is quite a lot going on in chess that is not captured by noughts & crosses. But because the two games share some important features, noughts & crosses can help improve our understanding of why skilled players tend to end the game with a draw.

Noughts & crosses is fun for small children. While the game between unskilled players tends to have a victor, after a bit of practice you quickly learn to reason via **backward induction**: you can figure out your opponent's response to your possible actions and take that into consideration before making your own move.

Once players learn to reason via backward induction, all noughts & crosses games are likely to end in a draw. In this way, noughts & crosses works as a simple model of chess, in which there are far more possible moves, but which, when played between skilled players is also likely to end in a draw.

Dealing with complexity: art and science

The primary concern of game theory is not board games like chess. Rather, its aim is to improve our understanding of interactions between people, companies, countries, animals, etc., when the actual problems are too complex to fully understand.

To do this in game theory we create very simplified models, which are called **games**. The creation of a useful model is both a science and an art. A good model is simple enough to allow us to fully understand the incentives motivating players. At the same time, it must capture important elements of reality, which involves creative insight and judgement to determine which elements are most relevant.

There is not one true model of any situation. There can be many models, each of which highlights a different aspect of the actual strategic interaction.

Rationality

Game theory usually assumes rationality and common knowledge of rationality. **Rationality** refers to players understanding the setup of the game and exercising the ability to reason.

Common knowledge of rationality is a more subtle requirement. Not only do we both have to be rational, but I have to know that you are rational. I also need a second level of knowledge: I have to know that you know that I am rational. I need a third level of knowledge as well: I have to know that you know that I know that you know I am rational. And so on to deeper and deeper levels. Common knowledge of rationality requires that we are able to continue this chain of knowledge indefinitely.

Keynes' Beauty Contest

The requirements for common knowledge of rationality are confusing to read. But worse, they might well break down in reality, especially in games with many players. A classic example is **Keynes' Beauty Contest**, in which English economist **John Maynard Keynes** (1883–1946) likens investment in financial markets to a newspaper competition in which readers have to choose the "prettiest face"; the readers who choose the most frequently chosen face win.

'It is not a case of choosing those which, to the best of one's judgement, are really the prettiest, nor even those which average opinion genuinely thinks the prettiest ... We devote our intelligences to anticipating what average opinion expects the average opinion to be.'

John Maynard Keynes

At first glance, Keynes' Beauty Contest has very little to do with financial markets: there are no prices, and there are no buyers or sellers. But they have one crucial feature in common. Success in financial markets depends on being one step ahead of the crowd. If you can *predict the behaviour of the average investor*, you can make a killing. Likewise in Keynes' Beauty Contest, if you can predict the average choice of newspaper readers, you can win the contest.

Thaler's Guessing Game

In 1997 the American behavioural economist **Richard Thaler** (b. 1945) ran an experiment in the *Financial Times*, a **Guessing Game** which was a version of Keynes' Beauty Contest.

GUESS THE NUMBER!

Readers pick a number between zero and 100. The winner is the contestant with the number closest to 2/3 of the average of all numbers entered in the contest.

Which number would you pick?

In Thaler's *Financial Times* experiment, the newspaper received more than a thousand entries. The entry 33 was the most frequently picked number, followed by the number 22. This suggests that many people reasoned one step and so chose 33. But many others thought that other readers would stop there and tried to be one step ahead of them by choosing 22 (which is 2/3 of 33).

If you believe that others will stop at the first step of reasoning, it is rational for you to stop at the second step.

Richard Thaler

However, if there is common knowledge of rationality, you know that others will not stop at the first step, so you can continue this **iterative reasoning** forever – a process of reasoning that involves repetition of the same process, taking the result from one round as a starting point for the next.

Game theorists solve the Guessing Game in a similar fashion using **iterative elimination of dominated strategies**.

Remember that you're looking for 2/3 of the average number entered into the contest. If all contestants were to pick the highest permissible number, 100, the average would be 100. Hence, no matter what one expects the average to be, it makes no sense to ever guess a number greater than 2/3 of 100, which is 67.

In other words, any strategy with a guess greater than 67 is **dominated** by 67. A strategy is dominated if it (in this case, a guess higher than 67) is worse than another strategy (guessing 67) regardless of what other players do. Hence, even if no one else is rational, all strategies with a guess greater than 67 can be eliminated.

If everyone else is rational, then each player can reason that no one would guess a number higher than 67. Hence, guesses above 45 (which is the closest integer to 2/3 of 67) are also eliminated. And because each player knows that the others know that everyone is rational they can each be certain that nobody else would choose a number greater than 45, and so they will not choose a number greater than 30 which is 2/3 of 45.

In the Guessing Game, iterative reasoning leads to smaller and smaller numbers, until all numbers above zero have been eliminated as dominated strategies. So, rational people with common knowledge of rationality would pick zero.

Problems with rationality and common knowledge of rationality

Zero, however, was not the winning number in the *Financial Times* experiment. The average number came out to 19 and so the winner had the entry 13.

The winning number was much higher than what game theorists would have predicted. Where does game theory go wrong? Does game theory not have any predictive power?

In this case, the assumptions of rationality and common knowledge of rationality are not satisfied. For instance, many contestants picked the number 100, which is not rational. Even if one were to mistakenly expect everybody to pick 100, the optimal response would be 67. These contestants either did not fully understand the rules of the game or they were not able to calculate 2/3 of 100.

The concept of rationality requires unlimited cognitive capabilities. Fully rational man knows the solutions of all mathematical problems and can immediately perform all computations, regardless of how difficult they are. Human behaviour is probably better approximated by **bounded rationality**. That is, human rationality is limited by the tractability of the decision problem (how easy it is to manage), the cognitive limitations of our minds, the time available in which to make the decision, and how important the decision is to us.

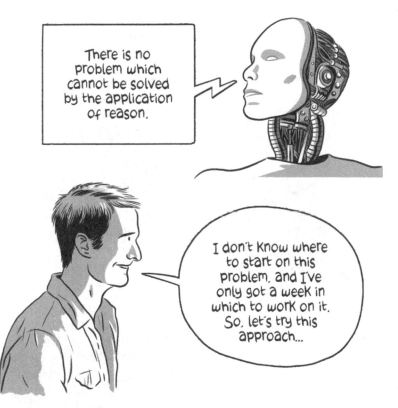

There is no problem which cannot be solved by the application of reason.

I don't know where to start on this problem, and I've only got a week in which to work on it. So, let's try this approach...

In addition to our bounded rationality, with a large number of participants, such as the Guessing Game received, it is hard to imagine that the common-knowledge-of-rationality assumption would hold. *Even if all players are rational*, you do not pick zero if you think that the other players do not know that you are rational. You would, therefore, make a guess higher than zero.

Do they know that I know that they know everybody else is rational?

If they don't, then they will pick a number greater than zero. So you should pick a number greater than zero.

Booms and crashes: applying rationality to financial markets

The Guessing Game and Keynes' Beauty Contest can explain the interesting fact that in financial markets we observe **bubbles** – excessively inflated prices – even if all participants are rational. This is because of a lack of common knowledge of rationality.

A fund manager may well be aware that the current stock price of a company does not reflect the true value of the company. However, it would be rational for him to buy the stock with the hopes of selling it at an even higher price in the future, if he expects that others anticipate the price to increase even further. The purchase decision would lead to increased prices of the stock today, creating a price bubble even if all traders are rational.

Simultaneous-move games

Often players do not know the actions of the other players when they make their own decisions. Games with this feature are called **simultaneous-move games**. In some cases the players are literally making their choices simultaneously (at the same time). In other cases, they might be making their choices at different times. But as long as they do not know what action the other players have chosen at the time they make their own decisions, we can treat them as moving simultaneously.

Consider this example: Rabbit Films has produced a spectacular superhero Christmas movie. It can release the film either in October or December.

One of Rabbit Films' big competitors, Weasel Studios, has produced a terrible movie at enormous expense. The romantic leads look like they can't stand each other and put on poor performances. Weasel also has the option of releasing its movie in either October or December.

More people go to the movies in December than October, which makes a December release attractive to both studios. But the two movies are aimed at the same demographic. If they both open in the same month they will steal audience from one another.

Each studio's revenue not only depends on its own release date, but also on the release date of the rival studio. Hence, the studios face **strategic interaction**. The payoff each studio gets from its choice of release date will depend on its rival's choice.

Strategic form of the game

We can analyze the game by listing the players' possible *actions* (release in October or in December) and *payoffs* (revenues) in a table, called the **strategic** (or **normal**) **form** of the game. The strategic form of a game is a table which is also known as a **payoff matrix**.

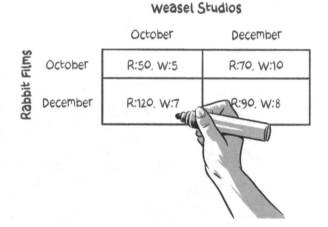

Weasel Studios

		October	December
Rabbit Films	October	R:50, W:5	R:70, W:10
	December	R:120, W:7	R:90, W:8

Each row represents one possible choice for Rabbit Films – October or December – and each column represents the possible choices for Weasel Studios. At the intersection of each row and column we put the payoffs for each player: in this example, the payoffs are the studios' revenues.

The matrix gives all possible outcomes of the game and specifies what each player would receive as a payoff in those outcomes. Both studios understand the payoff matrix and are aware that they both face the same matrix.

Payoffs

The meaning of payoff numbers varies depending on the problem being analyzed. In the movie release example, the payoff numbers are the revenue the movies are expected to generate (in millions of pounds) in each of the possible outcomes.

In other applications, the payoff numbers will have other interpretations. In biology, they are often the "fitness" of the player, where fitness is related to the chance of the animal reproducing and perpetuating the species. In many applications in economics, sociology, etc. the payoff numbers represent the relative "happiness" or "utility" of the players.

It may seem odd to assign numerical values to happiness or fitness. However, what matters for the players' decisions are not the numbers themselves, but rather how they relate to each other.

Preferences over outcomes are all that matter for the strategic interaction between the studios. All we need to know is which outcomes are better and which are worse for each of the players. The numbers are simply a convenient way to represent these preferences over outcomes.

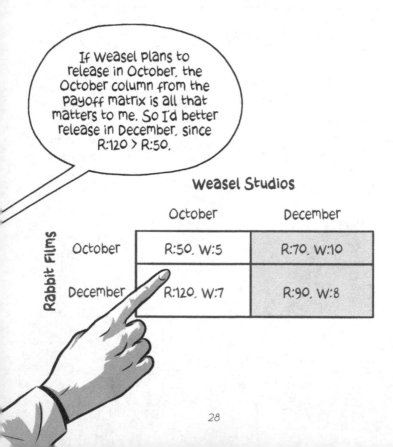

If Weasel plans to release in October, the October column from the payoff matrix is all that matters to me. So I'd better release in December, since R:120 > R:50.

Weasel Studios

Rabbit Films	October	December
October	R:50, W:5	R:70, W:10
December	R:120, W:7	R:90, W:8

There are, of course, many important situations where people care about the payoffs to others, as well as caring about their own payoffs. Families and friends may want to make each other happy. Divorcing couples and business rivals may want to hurt one another.

These situations can be easily analyzed via game theory by including *all* desires, both to look after ourselves and to help or hurt others, when writing down the payoffs. The payoff numbers in the table represent the *total payoff* that the players get from each outcome: in a particular outcome a player might benefit *directly*, as well as *indirectly* from hurting or helping others. The payoff numbers include everything they care about.

Once we write the game in strategic form, each player will only be concerned with increasing their payoffs.

Nash equilibrium

Now that we have formally specified the game by writing it in strategic form we can begin to think about what is likely to happen.

A fundamental concept in game theory is **Nash equilibrium**, named after the American mathematician **John Nash** (1928–2015). Nash didn't invent the idea of Nash equilibrium, which is much older, but he applied it to mathematical analysis of games in general, rather than just to specific examples, as had been done before.

We should expect each player to do the best he or she can, given what other players are doing.

John Nash

The idea of Nash equilibrium is both simple and powerful: in equilibrium each rational player chooses his or her **best response** to the choice of the other player. That is, he or she chooses the best action given what the other player is doing.

The best response of Rabbit Films

- If Rabbit expects Weasel to release in Oct, its best response is to release in Dec, since R:120 > R:50. Underline R:120.
- If Rabbit expects Weasel to release in Dec, its best response is to release in Dec, since R:90 > R:70. Underline R:90.

The best response of Weasel Studios

- If Weasel expects Rabbit to release in Oct, its best response is to release in Dec, since W:10 > W:5. Underline W:10.
- If Weasel expects Rabbit to release in Dec, its best response is to release in Dec, since W:8 > W:7. Underline W:8.

In equilibrium, both studios release in December. This is the only outcome where both studios have best responses to each other. If one studio is releasing in December, it is optimal for the other to release in December.

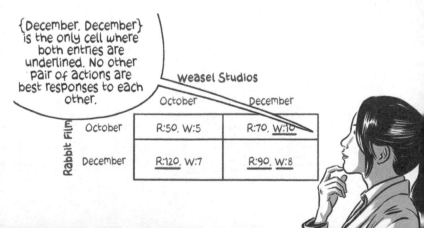

{December, December} is the only cell where both entries are underlined. No other pair of actions are best responses to each other.

Weasel Studios

		October	December
Rabbit Film	October	R:50, W:5	R:70, <u>W:10</u>
	December	<u>R:120</u>, W:7	<u>R:90</u>, <u>W:8</u>

One of the features of Nash equilibrium is that it is **regret free**. No studio would benefit if it were to deviate from its equilibrium strategy of releasing in December. The Nash equilibrium is also a **rational expectations** equilibrium. In equilibrium, Rabbit Films releases the movie in December with the expectation that Weasel Studios releases their movie in December. Indeed, Weasel Studios chooses December as their release date. Hence, the expectations are correct.

Prisoners' Dilemma

The most well-known game theory paradox is the Prisoners' Dilemma. The game was nicknamed by the Canadian mathematician **Albert Tucker** (1905–95). Professor Tucker's Prisoners' Dilemma Game is straight out of a Hollywood procedural crime drama where two prisoners are each offered a plea deal to rat on each other. The game illustrates the difficulty of acting together for common or mutual benefit given that people pursue self-interest.

The incentives that the Prisoners' Dilemma Game represents are common and have been useful in analyzing problems in a wide variety of areas, from competition between firms in economics to social norms in sociology to decision making in psychology to animals competing for scarce resources in biology, to computer systems competing for network bandwidth in engineering.

Alan and Ben get caught for jointly stealing a car. The police suspect that they were also involved in a hit-and-run but do not have evidence to convict them on this charge. The two prisoners are interrogated in separate rooms.

Alan and Ben each have two possible actions: to remain silent or to confess. Hence, there are four possible outcomes of the game:

> Alan is silent and Ben is silent
> Alan confesses and Ben is silent
> Alan is silent and Ben confesses
> Alan confesses and Ben confesses.

The decision is complicated by the fact that the jail time a prisoner serves not only depends on how he pleads but also on whether the other prisoner confesses or not.

Albert Tucker

This Prisoners' Dilemma can be represented in strategic form where each row in the payoff matrix represents a possible choice for Alan, and each column represents a possible choice for Ben. At the intersection of each row and column we put the payoffs for each player: in this case their jail time.

Ben

	Silent	Confess
Silent	A:-1, B:-1	A:-15, B:0
Confess	A:0, B:-15	A:-10, B:-10

Alan

If Alan and Ben are both silent, then they each go to jail for one year for grand theft auto. This is a bad thing, so their payoffs are negative (Alan:-1, Ben:-1). If both prisoners confess, each goes to jail for 10 years (A:-10, B:-10).

To obtain a confession on the hit-and-run, we offer a plea deal. If just one prisoner confesses and testifies against the other, he will walk free while the other goes to jail for 15 years.

The prisoners understand the payoff matrix and are aware that they both face the same matrix.

This is a simultaneous-move game: even if the prisoners' decisions are not literally simultaneous, we can think of them that way because the players are in separate rooms and so neither player knows the other's decision when making his own choice.

Notice that in putting the game into strategic form we do not say anything about what is likely to happen. We simply put down all potential outcomes, whether reasonable or not, and record the payoffs the players would get if that outcome occurred.

Now that we have written our problem in strategic form we can start to analyze what is likely to happen.

Clearly if Alan and Ben could come up with a joint response, they would both remain silent so that they could go to prison for only one year.

But this is not the equilibrium outcome. For Alan, the strategy "confess" **strictly dominates** the strategy "silent": it is always best to confess, no matter what he expects Ben to do.

If Ben confesses, it's best for me to confess since 10 years of jail is better than 15.
If Ben remains silent, it's still best for me to confess since walking out as a free man is better than going to jail for a year.

Similarly, whatever Ben expects Alan to do, Ben's best response is to confess.

In the Prisoners' Dilemma, both players confess in the Nash equilibrium. One standard way to write this outcome is:

{confess, confess}

This gives the choice of the row player (Alan) first, followed by the choice of the column player (Ben). In equilibrium both prisoners go to jail for 10 years.

Pareto efficiency

One interesting question to ask is whether the Nash equilibrium is **Pareto efficient** in the Prisoners' Dilemma Game. An outcome is Pareto efficient if there is no other potential outcome where somebody is better off and nobody is worse off. This notion of distributional efficiency is named after the Italian economist **Vilfredo Pareto** (1848–1923).

If an outcome is not Pareto efficient, it means that somebody could still be made better off without hurting anybody.

Vilfredo Pareto

The Nash equilibrium outcome of the Prisoners' Dilemma is not Pareto efficient because each prisoner would have been better off if both had remained silent; hence the nickname "Prisoners' Dilemma".

However, in many other games the Nash equilibrium is Pareto efficient. For instance, in the film studio game, there is no alternative outcome to the Nash equilibrium outcome that makes a studio better off without hurting the other side.

Network engineering

The incentives illustrated in the Prisoners' Dilemma Game are found in diverse situations. Indeed, once one starts viewing the world through this lens, it is difficult not to see a Prisoners' Dilemma everywhere.

For example, when wireless network routers, such as Wi-Fi routers or cell-phone towers, use the same frequency and are within range of each other, they interfere with each other's communication, slowing down the speed of both routers.

One solution to this problem is to lower the transmission power of both routers so that they are no longer within range of one another. But if only one router is using low power, then its signal is overwhelmed by the high-power router.

The network router situation can be represented by this payoff matrix.

Router B

	High Power	Low Power
High Power	A:5, B:5	A:15, B:2
Low Power	A:2, B:15	A:10, B:10

Router A

The engineers of each router have to decide whether to broadcast at high or low power, and the payoffs are the data transmission speeds in millions of bits per second (Mbps). In this game, broadcasting at higher power gives a router an advantage at the expense of the other router, just as confessing did in the Prisoners' Dilemma Game.

Each router finds that broadcasting at high power gives the fastest speed regardless of what the other router is doing; "high power" is a dominant strategy. In Nash equilibrium both routers broadcast at high power and achieve a data transmission speed of 5Mbps each – just as in the Prisoners' Dilemma both prisoners confess and go to jail for a long time.

Service is slow!

If both routers operated at "low power", they would have a transmission speed of 10 Mbps each. Nevertheless, when their power is set independently, neither router will pick low power because each individual router can do better by increasing its broadcast power.

If both routers are part of the same network then it is possible to force them both to use low-power mode to minimize the conflict. Most routers have "advanced" settings, which are there to force them to cooperate with other routers on the network rather than to compete aggressively for resources. The advanced settings are there to help the network administrator override this kind of Prisoners' Dilemma problem.

The tragedy of the commons

The network router problem is closely related to the **tragedy of the commons**, a concept conceived by **William Forster Lloyd** (1794–1852), much earlier than the conception of the Prisoners' Dilemma. In an essay about overgrazing of cattle, Lloyd argued that farmers may be acting in self-interest, contrary to the best interest of the whole group, and depleting the feeding potential of the common land.

In the economics literature, the term "commons" has evolved to encompass any shared resource. So, in the network router problem the commons is the wireless bandwidth that the routers are competing over. In this example, the overuse of the resource does not cause long-term damage or depletion of a natural resource, as it does in Lloyd's overgrazing example. Nevertheless, the individual incentives to overuse the resource at the expense of the group are the same.

Nuclear build-up

The Prisoners' Dilemma Game was originally conceived by mathematicians **Melvin Dresher** (1911–92) and **Merrill Flood** (1908–91) in 1950 while working on a US Air Force project. The goal at the time was to further our understanding of global nuclear strategy.

In Dresher and Flood's original formulation of the Prisoners' Dilemma, the two players are the United States and the USSR (although by the height of the Cold War in the 1980s the number of players had expanded considerably). Each country must decide whether to increase its nuclear arsenal. If a country doesn't add to its arsenal, it saves the cost and implicit risk of accident. But each country has an incentive to increase its arsenal to enhance its geopolitical position. It is in each country's self-interest to invest in nuclear weapons, no matter what the other county is doing. The Nash equilibrium of the game, therefore, is global nuclear build-up.

'A nuclear war cannot be won and must never be fought. The only value in our two nations possessing nuclear weapons is to make sure they will never be used. But then would it not be better to do away with them entirely?'

US President Ronald Reagan, 1984 State of the Union speech

44

'A world without nuclear weapons may be a dream, but you cannot base a sure defence on dreams.'

British Prime Minister Margaret Thatcher, 1987

The Nash equilibrium outcome in the nuclear arms race is not Pareto efficient because both countries would be better off if neither engaged in nuclear build-up. However, as Dresher and Flood argued, this couldn't be an equilibrium. If the USA were to have stopped nuclear build-up, the USSR would have continued building its own arsenal to capture the "super-power" position. And it would not have been rational for the USA to stop nuclear build-up in the first place.

Cooperation

In the Prisoners' Dilemma, although there is benefit to **cooperative behaviour**, individual incentives encourage conflict. In the network engineering example, it is possible to overcome this problem if one person controls both routers. But in human interaction achieving cooperation can be more difficult.

Social psychologists study conflict and cooperation to understand the manner in which individuals' behaviours are influenced by social groups. Consider an example of the Prisoners' Dilemma known as the Roommate Game. This game both emphasises the theory's broad applications and provides a framework to think about how social norms can help to overcome individual incentives for excessive conflict.

46

Alice and Beth share an apartment. They like a clean kitchen, but neither of them enjoys doing the dishes. Each girl has the choice to clean up or not. They are involved in strategic interaction since Alice's happiness (payoff) is affected by Beth's choice of action, and vice versa.

If neither roommate cleans up, Alice has payoff of 10 (A:10), as does Beth (B:10); these "happiness" payoff numbers just serve to show us which of the outcomes each girl prefers. If only Beth cleans up, Alice's payoff goes up to 20, but doing the dishes reduces Beth's payoff to 8 (A:20, B:8). If only Alice cleans up, the reverse is true (A:8, B:20). If they share the job, the burden of cleaning up is halved and each gets a payoff of 14. Beth and Alice are well aware of how each outcome will affect their happiness.

	Beth	
	Don't clean up	Clean up
Alice Don't clean up	A: 10, B: 10	A: 20, B: 8
Clean up	A: 8, B: 20	A: 14, B: 14

The Nash equilibrium of the game is {don't clean up, don't clean up} because if either expects her roommate not to clean up, the best response is not to clean up.

There is a **free-rider problem** in the Roommate Game. Alice has the highest payoff when she relaxes while Beth does the dishes. The same goes for Beth.

So, the girls have a messy kitchen in equilibrium with payoffs of 10 each. If they were to cooperate they would have a greater payoff of 14 each. Nevertheless, maintaining a clean kitchen through cooperation is not an equilibrium outcome. The moment one expects the other to clean up, the incentive to free-ride rises to the surface.

She's such a free-rider. I would have felt better if I'd just left the dirty dishes and watched TV.

Education

One way out of the free-rider problem is to change the payoffs in the payoff matrix. Early childhood parental involvement and schooling can impose a moral cost when engaged in non-cooperative behaviour (such as leaving dishes in the sink).

I feel really guilty when I see dirty dishes piling up in the sink.

The imposition of a moral cost may initially seem bad for the girls. After all, who likes to feel guilty? But in their social interaction, a moral cost can change the equilibrium by encouraging both girls to behave more cooperatively. Alice and Beth can be better off if both have moral values because it allows them to achieve the benefits of cooperation that were unattainable before.

Assume that there is a moral cost of not doing your share. If either girl doesn't clean up, she feels guilty and her happiness goes down by 7 in the Roommate Game. The Nash equilibrium now would be {clean up, clean up} with payoffs of 14 for each girl.

Since players choose to cooperate, in equilibrium they do not pay the moral cost. There is a **Pareto improvement** in the outcome of the Roommate Game due to moral values; players' equilibrium payoffs go up from 10 to 14.

Beth

	Don't clean up	Clean up
Don't clean up	~~A: 10, B: 10~~ A: 3, B: 3	~~A: 20~~, B: 8 A: 13
Clean up	A: 8, ~~B: 20~~ B: 13	A: 14, B: 14

Alice

That wasn't so hard!

And I didn't feel guilty because we ended up cooperating.

Environmental policy and cooperation

International cooperation on environmental protection is like the Roommate Game writ large. Each country prefers to remain passive while the others adopt costly abatement technologies to reduce CO2 emissions. A way out of this free-rider problem is to sign an international treaty that legally commits countries to pay monetary fines if their CO2 emissions are in excess of mutually agreed limits. However, it has been difficult to get the major polluters – China, India and the USA – to ratify such an international treaty.

Why is it so hard to secure an international agreement on curbing emissions when it is beneficial for all? .

While cooperation would be good for all, the outcome most preferred by the USA, is for other countries to sign an international treaty with monetary fines while the USA does not. It is always sweet to get a free ride.

One way to view environmental activism is as an effort to change the social norms. Political pressure can impose a cost on politicians who do not support environmentally friendly policies. This can change the payoffs of the nations' policymakers, just like the moral cost of guilt changed the payoffs in the Roommate Game. Political pressure can potentially lead to a better outcome if it creates an equilibrium where countries cooperate.

Multiplicity of equilibria

So far we've looked at games with a single Nash equilibrium. In these games, the Nash equilibrium gives a single prediction for the players' behaviour. However, people frequently find themselves in environments with many Nash equilibria. In games with multiple Nash equilibria, the concept of Nash equilibrium by itself does not provide us with sufficient tools to predict what will happen.

When there are many equilibria, which one will the players actually play?

In search of the answer to this question, the 2005 Nobel Prize-winner, American economist and professor of foreign policy **Thomas Schelling** (b. 1921) redefined the scope of economics and its place in the social sciences.

Multiplicity of equilibria: Battle of the Sexes

The classic Battle of the Sexes Game provides a clear understanding of the incentives in a game with multiple Nash equilibria. The game might seem pretty mundane and based on outdated stereotypes, but it's a useful illustration because the same form of incentives appear in many different situations.

At breakfast, Amy and Bob, a couple, decide to spend the evening together, but each wants to partake in a different activity. They agree to call each other during the day and make a decision then about where to go.

	Bob	
	Football	Dance class
Football	A: 5, B: 10	A: 0, B: 0
Dance class	A: 0, B: 0	A: 10, B: 5

Amy (left row label)

The matrix gives the "happiness" payoffs. These numbers simply serve to show us which of the outcomes each player prefers. For example, if Amy and Bob go to the football match, Amy gets a payoff of 5 (A:5). If they both go dancing, Amy gets a payoff of 10 (A:10). The exact numbers are unimportant, they just serve as a shorthand to tell us that she would rather go dancing together than to football together, since 10>5.

While Amy and Bob have different preferences over their top choice of activity, they love spending time together. They both think that the worst alternative would be spending the evening alone. If they end up going to different activities, they each get a payoff of zero.

I like football, but mostly I just want to be with Amy.

During the day, phone service breaks down. Amy and Bob need to make a decision about where to go without communicating with each other and without having observed each other's decision. So, this is a simultaneous-move game.

It is a Nash equilibrium for both to go to the football match.

I think Bob will go to the football match. Therefore, I'll go to the match.

I think Amy will expect me to be at the football match. Therefore she will go to the football match. So I shall go to the match.

However, it is also a Nash equilibrium for both to go dancing.

I think Bob will expect me to be at the dance class. Hence he will be at dance class. So I will go dancing.

I think Amy will be at the dance class. Therefore I will go dancing.

In the Battle of the Sexes Game, there are two Nash equilibria in which the players choose a particular activity with certainty: the football equilibrium and the dancing equilibrium.

But what will Amy and Bob end up doing?

'The existence of multiple equilibria is a pervasive fact of life that needs to be appreciated and understood, not ignored.'

It is plausible that the couple in the Battle of Sexes Game ends up with **coordination failure** due to misaligned expectations. In this case the game theorist would observe an "out-of-equilibrium" outcome where the couple spends the evening separated: neither of the two possible Nash equilibria come to pass.

I expected Bob to be at the football match. But he's not here! He must have thought I would go dancing.

I thought Amy would go dancing. But she isn't here! She must have expected me to go to the match.

There are possible ways to avoid coordination failure in games with more than one equilibrium ...

Social norms

In environments with multiple equilibria, players may coordinate their expectation on one equilibrium using social norms. For instance, if Bob tends to get his way in the relationship, both Amy and Bob would presume that Bob's preferred equilibrium will prevail whenever there is multiplicity of equilibria. In this case, not only would Bob be happy to go to the match with Amy, but also Amy would be happy because she is with Bob rather than spending the night alone.

While the Battle of Sexes Game does not give conditions under which societies evolve into **patriarchy** (a society structured around male advantage), it does give an insight into a potential benefit of gender-based dominance. This may be one of the reasons why it is so difficult to move society to a fairer system.

When a game has more than one equilibrium, the environment or the history of the game may focus the players' expectations on a particular equilibrium, in which case their rational response would be to play it.

This **focal-point effect** means that culture and history can affect our rational behaviour.

Coordination devices

In games with multiple equilibria, if a social norm is not present, players may use a **coordination device**, a shared observation or common history to help coordinate expectations on the same equilibrium.

For instance, the dance studio may heavily advertise on the radio station that Amy and Bob listen to. It is rational for the dance studio to invest in advertising if the studio expects advertising to coordinate consumer expectations about which equilibrium will be selected. Amy and Bob can reason that the studio advertises because listeners use advertising to coordinate expectations. Therefore, in the absence of direct communication, they may use the advertising they hear during the day as a way to coordinate their expectations on the dance class equilibrium.

Banking and expectations: bank runs

The way that banks make money is by taking our deposits and loaning some of that money to businesses and consumers who pay interest to the bank. This is good for the banks, and it also allows people to buy homes and enables businesses to invest. But it means that not everyone can take out their deposits at the same time. Most of the money has been loaned out and won't be available until all the mortgages are paid off.

Hence, no matter how healthy the financial state of a bank may be, any bank will go under if faced with a **bank run** (where everybody tries to withdraw their money at the same time).

As in the Battle of the Sexes Game, in banking there are multiple Nash equilibria. Depending on people's expectations, we may observe business as usual or a bank run.

If depositors expect that others will not be withdrawing their money, they wait until their deposit matures to collect their interest earnings. However, there is a second Nash equilibrium; if depositors anticipate that other depositors will withdraw their money early, each individual depositor will rush to the bank and withdraw his or her own money before the teller closes the window.

The belief that there will be a bank run is a **self-fulfilling expectation**: the expectation itself causes the bank run.

One of the main functions of a central bank is to reduce the chance of self-fulfilling runs on banks.

In most industrialized countries, central banks take the role of the **lender of last resort**: they stand ready to loan money to a bank to get it through expectations-driven bank runs.

Additionally, deposit insurance is provided for small depositors so that everybody is guaranteed to get their money back even if the bank goes under.

Therefore, people do not need to rush to withdraw their deposits even if they expect others to withdraw their money.

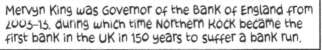

Mervyn King was Governor of the Bank of England from 2003-13, during which time Northern Rock became the first bank in the UK in 150 years to suffer a bank run.

However, even with a lender of last resort and deposit insurance, bank runs are not entirely avoidable. Individual depositors can reasonably presume that there would be a time delay after a bank run before the deposit insurance pays out.

Since there are always multiple equilibria in banking, people's expectations about what will happen determine the outcome. Even positive statements or actions by bankers or policymakers can backfire if people take them as a sign of weakness.

Mixed-strategy Nash equilibrium

So far we have looked at games that have a **pure-strategy Nash equilibrium**. This is an equilibrium in which players pick a particular choice with certainty. But not all games have such an equilibrium. Remember the Rock-Paper-Scissors game you played as a kid? "Scissors" cut paper, "rock" breaks "scissors", and "paper" wraps around "rock". This is a **zero-sum game**: if one person wins, the other loses.

What makes this game fun for kids is the *unpredictability* of the outcome. What makes it interesting for game theory is that it has no equilibrium where players behave predictably. If a player is predictable, the other player will exploit that and win. So, players try to be unpredictable; the game does not have a pure-strategy Nash equilibrium.

While Rock-Paper-Scissors does not have a pure-strategy Nash equilibrium, it does have a **mixed-strategy Nash equilibrium**. This means that in equilibrium the players randomize over possible pure strategies: "rock", "paper" and "scissors".

In order to win I have to be unpredictable.

However, not every random play is a mixed-strategy Nash equilibrium. Just randomizing is not enough; the mixed strategies of the players need to be best responses to each other to form a Nash equilibrium.

Let's look at a strategy that *cannot* be sustained in Nash equilibrium.

Suppose that Jack's strategy is to announce paper with 10% chance, rock with 80% chance, and scissors with 10% chance.

Susan's best response to Jack's strategy is to announce paper with certainty, which would give her an 80% chance of winning, the probability that Jack announces rock.

Why do you win most of the time even though I'm playing a randomized mixed strategy?!

Although you are randomizing, you are choosing rock so often that I'm very likely to win just by choosing paper.

Jack and Susan's strategies are not a Nash equilibrium: the players' choices are not best responses to each other. Given Susan's strategy, it would be Jack's best response to pick scissors with certainty instead of his randomized strategy.

The Rock-Paper-Scissors Game has just one equilibrium: each player plays a mixed strategy of choosing each of the three possible choices, (rock, paper or scissors) with equal probability.

Jack's randomization with even probabilities means that Susan has no preference over her three possible choices. If Susan plays scissors she has a 1/3 chance of winning (which happens if Jack plays paper), a 1/3 chance of losing (which happens if Jack plays rock) and a 1/3 chance of a tie (which happens if Jack also plays scissors). But any other choice besides scissors would also give her the same payoffs.

I am **indifferent** between rock, paper and scissors. Because each choice gives me the same payoff, I am willing to randomize in my decision.

The same reasoning is true for Jack. As long as Susan is choosing each of the options with even probability, Jack gets the same expected payoff from choosing any of his three options. So he is willing to randomize.

The Currency Speculation Game

Mixed-strategy Nash equilibrium has applications in a wide variety of fields. It can capture the spirit of *surprise* in games where players are unpredictable. For instance, it can improve our understanding of speculative attacks, which are typically *sudden* and *unexpected*.

On "Black Wednesday" – 16 September 1992 – in a sudden **speculative attack** investors sold massive amounts of British pounds with the anticipation of a *devaluation* (a drop in the value of the pound against other currencies). At the time, the value of pound was fixed by the Bank of England against other European Union currencies. On the day, the Bank bought £4 billion to keep the pound from losing value.

However, unable to resist the market forces, the next day the Bank let the value of the pound fall by more than 10%. Speculators who had sold pounds and purchased German marks just one day earlier made huge profits. The Bank made huge losses. One of the large speculators, Hungarian-American businessman **George Soros** (b. 1930) became known as "the man who broke the Bank of England".

How to make sense of Black Wednesday? Why didn't the Bank of England devalue the pound a day before the attack to avoid massive losses?

For the investor, it is best to be unpredictable about when to engage in a speculative attack. If the central bank could predict the attack, the bank would pre-emptively devalue its currency the day before the attack to avoid losses. The speculators would then be too late to take advantage of the devaluation.

'The financial markets generally are unpredictable... The idea that you can actually predict what's going to happen contradicts my way of looking at the market.'

George Soros

There is no pure-strategy Nash equilibrium in the Currency Speculation Game. Just like in the Rock-Paper-Scissors Game, the only equilibrium in the Speculation Game is in mixed strategies. Speculators randomize the timing of the attack so that the central bank cannot predict the exact date of the speculative attack. This explains why the Bank of England was unable to pre-empt the speculative attack on Black Wednesday.

The Chicken Game

Mixed-strategy Nash equilibrium is intuitively appealing when there is no pure-strategy Nash equilibrium because players choose to be unpredictable. Mixed-strategy Nash equilibrium is also interesting in environments with multiple pure-strategy Nash equilibria where each player prefers a different equilibrium outcome.

A classic example is the **Chicken Game**: two teenagers drive towards each other in a contest of courage to see who will keep driving longer. There is a pure-strategy Nash equilibrium where one teenager continues straight ahead while the other turns off, and there is another equilibrium where their roles are reversed. Of course, each teenager prefers the equilibrium where they are the "rebel" and the other is the "chicken".

The Chicken Game is thrilling because there may be a crash if neither teenager chickens out. However, a crash is not a possible equilibrium outcome if we only focus on pure-strategy Nash equilibria. Obviously, if one is continuing straight ahead, the best response of the other would be to turn off, avoiding the crash.

To capture the thrill of the Chicken Game, we need to consider the mixed-strategy equilibrium where both players randomize between continuing straight ahead and turning off. In the mixed-strategy Nash equilibrium, a head-on collision is one of the possible equilibrium outcomes.

They both want to be the rebel. What if nobody chickens out? What is the chance that they survive this madness?

The Exit Game

An economic application of the Chicken Game is the Exit Game. This game provides a clear illustration of how to find equilibrium probabilities in the mixed-strategy Nash equilibrium.

Smallville has two grocery stores, Kalemart and Carrotco. Recently the town's population decreased significantly. The town is now too small to allow Kalemart and Carrotco to make money if they both continue to operate in town. However, either business can be profitable if only a single company operates. Hence, each company prefers that the other exits the market, while it stays and enjoys the monopoly position in Smallville.

This payoff matrix gives the profit and loss to Kalemart and Carrotco for each possible outcome. If both Kalemart (K) and Carrotco (C) stay in town, each business makes a loss, K:-20 and C:-50. These fictional figures are shorthand for more realistic profit figures, such as -£20,000 and -£50,000. If both exit the market, each has zero profit.

If Kalemart stays while Carrotco exits the market, Kalemart makes a profit of K:80, while Carrotco's profit is C:0. If Carrotco stays and captures the monopoly position, Carrotco receives a profit of C:100, while the profit to Kalemart is K:0.

CarrotCo

		Stay	Exit
Kalemart	Stay	K:-20, C:-50	K:80, C: 0
	Exit	K:0, C: 100	K:0, C: 0

We've got two pure-strategy Nash equilibria: Kalemart stays and Carrotco exits, or Carrotco stays and Kalemart exits.

The firms are likely to struggle over their preferred position in the Smallville market since each wants to be the only grocer in town. The mixed-strategy Nash equilibrium reflects the spirit of this struggle. No grocery store wants to give up the fight, just as no teenager wants to be called "chicken" in the Chicken Game. And just as in the Chicken Game, where each teenager has a chance of acting as the rebel, in the Exit Game each business stays in Smallville with some probability, but not with certainty.

In the Rock-Paper-Scissors Game, players announce rock, paper or scissors with 1/3 chance. What about the Exit Game in Smallville? What is the equilibrium probability that Carrotco stays? How about Kalemart?

The key to solving for the probabilities in the mixed-strategy Nash equilibrium is to realize that a business *randomizes* between "stay" and "exit" only if it is *indifferent* between these actions. And the business is indifferent if its expected profit from "stay" is equal to its expected profit from "exit".

If the expected profit from one action were higher than the other, the business would prefer the action with the higher expected profit, and it would choose that action with certainty. In equilibrium, there is randomization and hence *uncertainty* about a store's action only when the store is indifferent, when its expected profit is the same from either action.

SORRY WE'RE CLOSED

What should I do? There are compelling reasons both for staying in town and for going.

If Carrotco exits, its expected profit is zero no matter what Kalemart does:

$$\text{Expected profit if ``exit''} = 0$$

On the other hand, if Carrotco stays, its expected profit depends on the probability with which Kalemart stays. Let k stand for the probability that Kalemart stays. If $k = 0$, there is no chance that Kalemart stays. If $k = 1/2$, then there is 50% chance that Kalemart stays. And $k = 1$, means that Kalemart stays with 100% chance. $(1 - k)$ is the probability that Kalemart exits.

If Carrotco stays, it makes a loss of −50 if Kalemart also stays, which happens with probability k. Carrotco gets 100 if Kalemart exits, which happens with probability $(1-k)$. Therefore for Carrotco:

$$\text{Expected profit if ``stay''} = \underbrace{-50(k)}_{\substack{\text{profit if Kalemart stays} \\ \text{* chance Kalemart stays}}} + \underbrace{100(1 - k)}_{\substack{\text{profit if Kalemart exits} \\ \text{* chance Kalemart exits}}}$$

Carrotco is indifferent between "stay" and "exit" if its expected profits from the two choices are equal:

$$\begin{array}{ccc} \text{Carrotco's} & = & \text{Carrotco's expected} \\ \text{profit if it exits} & & \text{profit if it stays} \end{array}$$

$$0 = -50(k) + 100(1-k)$$

To find Kalemart's equilibrium chance of staying, solve this for k, which gives $k = 2/3$.

I am indifferent between "stay" and "exit" because Kalemart will stay in business with 2/3 chance and will leave with 1/3 chance.

Carrotco's equilibrium chance of staying is 4/5. This can be calculated in a similar fashion by looking for the probability that makes Kalemart indifferent between staying and exiting.

If Kalemart stays, it makes a loss of -20 if Carrotco stays, which happens with probability 4/5. Kalemart gets a profit of 80 if Carrotco exits, which happens with 1/5 probability. For Kalemart, in equilibrium the expected profit from exiting (which is zero) is the same as the expected profit from staying.

$$0 = -20(4/5) + 80(1/5)$$

Kalemart's profit if it exits

Kalemart's expected profit if it stays

I really want to be the only one left in this town. But Carrotco is so likely to stay that staying in town gives me the same expected payoff as leaving.

SWEET SEEDLESS SPANISH SATSUMAS
35 P Pound

SOUND FIRM SPANISH SALAD TOMATOES
20 P Pound

Since in equilibrium Kalemart stays with probability 2/3 and Carrotco stays with probability 4/5, we can calculate the probabilities of each possible outcome in Smallville.

Both firms exit the Smallville market with a 1/15 chance, which is the probability that Kalemart exits multiplied by the probability that Carrotco exits, (1/3) * (1/5) = 1/15.

Both stores stay open for business with an 8/15 chance, which is the probability that Kalemart stays multiplied by the probability that Carrotco stays, (2/3) * (4/5) = 8/15. In this case both stores make a loss. This outcome is similar to the one in the Chicken Game, in which both teenagers act as rebels and die in a car crash.

I knew that you might stay open. I took a chance, but now I'm broke.

The chance that Kalemart ends up being the only store that stays open for business and the chance that Carrotco captures the monopoly position can be calculated in a similar manner.

It is also possible to create a version of the Exit Game where if both players stay in town, they still have the option of exiting at a later date. In that case the struggle can last a while, with huge losses accumulating over time. This is known as a **war of attrition**. The term is borrowed from military strategy. Long and damaging fights can occur in these types of games even when the prize may be small in relation to the accumulated costs.

Criticism and defence of mixed strategies

Of all topics in game theory, mixed-strategy Nash equilibrium probably elicits the strongest feelings. Proponents of mixed strategies point out that many games, such as the Rock-Paper-Scissors Game or the Currency Speculation Game, do not have any pure-strategy Nash equilibria, but they do have an interesting mixed-strategy Nash equilibrium. They also point out that even in games like the Chicken Game or the Exit Game, which do have pure-strategy Nash equilibria, the mixed-strategy Nash equilibrium is often the most intuitive since it can capture uncertainty in these environments.

However, critics of mixed strategies argue that randomization is not a reasonable description of human behaviour. Do people really randomize when making a decision? Furthermore, since players are indifferent between different actions in equilibrium, what motivates them to choose the exact probabilities that make other players just indifferent?

One powerful defence of mixed strategies is the **"purification"** interpretation of mixed-strategy Nash equilibrium. This was developed by Hungarian-American economist **John Charles Harsanyi** (1920–2000), who shared the 1994 Nobel Prize in Economics with John Nash and German economist **Reinhard Selten** (b. 1930).

Harsanyi points out that even if players play pure strategies, if they are slightly uncertain about each other's payoffs, from the outside they will seem as if they are randomizing between actions.

Harsanyi's remarkable "purification" argument proves that if the players are almost, but not quite, certain about each other's payoffs, from the individual's point of view the other player's chance of choosing a particular action is *exactly* the probability we get in a mixed-strategy Nash equilibrium without uncertainty about payoffs.

This means that the mixed-strategy Nash equilibrium is relevant even if you don't believe that it is human nature to randomize when making decisions.

Tax evasion

Players randomizing over possible actions is one interpretation of the mixed-strategy Nash equilibrium. Slight uncertainty about other players' payoffs is a second interpretation. The mixed-strategy Nash equilibrium can also be interpreted in a third way, as illustrated in the Tax Evasion Game between taxpayers and the tax authority.

Consider a business taxpayer who must declare how much tax she owes. For simplicity, suppose she has two choices: complying with the tax law, or evading taxes. Suppose that there are no moral implications of tax evasion.

If I am going to be audited for sure, I would rather comply with the tax law. If there is no chance of an audit, I prefer to evade taxes.

The tax authority can catch a tax-evader for sure if it pays for an expensive audit. But the audit is useless if the taxpayer is not evading.

There is no pure-strategy Nash equilibrium in the Tax Evasion Game.

A citizen would comply for sure if an audit is certain. This cannot be a Nash equilibrium: if the taxpayer is complying for sure, then there is no need for the government to audit.

A citizen would evade taxes for sure if she is certain that there will be no audit. Clearly this cannot be a Nash equilibrium either: if the taxpayer is evading, then the tax collector would rather audit.

The only equilibrium is in mixed strategies: taxpayers randomize between compliance and evasion, and the tax collector randomizes between auditing and not.

We don't audit everyone's taxes – so the probability that you get audited is less than 100%, but I can assure you that the probability is greater than zero.

If the Tax Evasion Game is being played by many citizens, a compelling alternative interpretation of the mixed-strategy Nash equilibrium is that each individual citizen plays a pure strategy, "comply" or "evade", but the mixed-strategy Nash equilibrium probabilities give the fraction of all citizens who play the pure strategy "comply" and the fraction of all citizens who play the pure strategy "evade". The tax collector knows the ratio of tax-evaders to taxpayers in compliance but doesn't know who is a complier and who is an evader.

Even though I am not an evader, I am being audited. Because there are people out there who evade their taxes, from the government's point of view, it is just random chance that I might be evading taxes.

Repeated interaction

Back in 1883, the French economist **Joseph Louis François Bertrand** (1822–1900) studied price competition between a few firms selling identical products. In his analysis, the incentives firms face are similar in spirit to the incentives in the Prisoners' Dilemma Game.

It is in each firm's best interest to "undercut" the price of the other to capture the whole market. In equilibrium, firms make low profit. If they were to collude at a high price, each could make handsome profits.

Bertrand predicted that firms will undercut one another in equilibrium, similar to the {confess, confess} outcome in the Prisoner's Dilemma. Despite this prediction, in markets with a small number of firms we often observe high collusive prices. Most Western democracies have "antitrust" regulation to avoid this type of **collusion** (cooperation among firms) and to promote competition.

To understand when players collude in a Prisoners' Dilemma type of situation, we need to move beyond **one-shot games** (where players play the game only once and then the game ends) and to start thinking about more realistic settings with **repeated interaction**, where players play the same game again and again.

Would we observe cooperation in equilibrium in the Prisoners' Dilemma if players interact repeatedly?

Imagine that both players know that they will play the Prisoners' Dilemma Game not once but twice. To find the equilibrium of the game with repeated interaction, we first predict the equilibrium of the game in the *last* round. And then we reason what the equilibrium would be in the *first* round. This line of reasoning is called **backward induction**.

At the end of the game

In the second round, players know that it's the last round, so there is no need to try to change the future outcome. Hence, the last round of the game is just a one-shot Prisoners' Dilemma: nobody cooperates.

Players can reason that there will be no cooperation in the second round no matter what happens in the first round. Hence, from the players' viewpoint, the first round of the game is no different from a one-shot Prisoners' Dilemma either. So, in equilibrium there is no cooperation at any stage of the game.

In fact, even if the Prisoners' Dilemma Game were repeated over many rounds, we would never observe cooperation in any round as long as the game has a certain final round. Backward induction unravels the game from the last round.

What if there is no definite last stage?

Israeli-American mathematician **Robert John Aumann** (b. 1930), who shared the 2005 Nobel Prize in Economics with Thomas Shelling, studied cooperation as an equilibrium outcome when a game has an **infinite horizon**, which means that the game is repeated forever. With an infinite horizon, backward induction does not unravel cooperation from the last round, since there is no certain last round.

The first condition for cooperation to be an equilibrium outcome is that players' strategies have an element of punishment for past bad behaviour (non-cooperative actions). To avoid future punishment, players may choose to be cooperative.

'In **continuingly** competitive games, individual self-interest can dictate a kind of cooperative behaviour sustained due to the fear of punishment by the other players for failing to cooperate.'

Robert John Aumann

In an infinite-horizon Prisoners' Dilemma Game, where the game is played repeatedly, forever, consider the so-called **grim strategy**: the player starts out with a cooperative action (depending on the game, this may be a prisoner keeping silent, a roommate doing the dishes, or a firm setting a high, collusive price). In subsequent rounds, the player cooperates if the other player has always cooperated. But the player **defects** (a prisoner confesses, a roommate stops doing the dishes, or a firm sets a price lower than the collusive price) if the other player has ever defected in the past.

How do you and your rival manage to keep your prices high instead of engaging in cut-throat competition?

We cooperate because we are both scared of what would happen otherwise.

Both players playing the grim strategy can be a Nash equilibrium in a repeated Prisoners' Dilemma type of game if players are *patient* enough (if they are able to resist the temptation of a high payoff today in order to be able to collect cooperative payoffs in the future). In this case, punishment for defection can deter the players from non-cooperative actions.

However, if players are *impatient*, they will be tempted to defect today despite the punishment in the future. Knowing this, the rival would not behave cooperatively in the first place. So, with impatient players cooperation cannot be sustained in equilibrium.

> I know my competitor. He needs money urgently! He will undercut my price this quarter and make a huge profit by stealing my customers. I cannot allow that. I will set a low price as well to be able to keep my customers.

When players are patient, for the threat of punishment to be a *deterrent* to defection, the threat needs to be *credible*. The grim strategy may not be credible if the player who punishes also receives low payoffs due to the punishment. Hence, if collusion breaks down, both players have an incentive to **renegotiate**, ignore the deviation and simply start colluding all over again. However, if players expect that they can renegotiate rapidly, collusion is not sustainable to start with.

However, if players expect renegotiation to take time, then the threat can have a deterrent effect and produce a collusive outcome in equilibrium.

Even if the game is not repeated forever, if the players are uncertain about when the game ends, cooperation can be sustained in equilibrium as long as they believe there is a good chance that the game continues to the next round. If so, there is a good chance that defections will be punished in the future, so cooperation can be maintained.

However, if there is a high chance that the game ends in the next stage, a player acts opportunistically and defects to capture high payoffs in the current round. But knowing this, their rival does not behave cooperatively either. Collusion does not take place.

I don't know how much longer I can keep my position at the company. I will sell as much as I can this year, even if that lowers prices and pisses off my competitors.
After all, I might not even be here next year to suffer the consequences.

FINISH

Prisoners' Dilemma experiment

One of the founding fathers of experimental economics, **Reinhard Selten** (who shared the 1994 Nobel Prize in Economics with John Nash and John Charles Harsanyi), conducted an experiment with participants who played a version of the repeated Prisoners' Dilemma Game for real money. Players were uncertain about the number of repetitions, but they knew that the experiment wouldn't last longer than a certain amount of time.

The experiment's results were broadly in line with game theory. Cooperative outcomes were often recorded as long as the end of the game was not in sight. But as time was running out and the end of the game drew near, players started to defect and mutual coordination broke down.

Evolutionary game theory

Most of game theory involves thinking of people, companies or countries as making rational decisions. It then investigates what decisions they make when they interact with others who also make rational decisions.

However, many behavioural economists and biologists, such as British evolutionary biologist **John Maynard Smith** (1920–2004) and American evolutionary biologist **George Price** (1922–75), look at interaction from another perspective. They often think of people or animals as being socially or genetically programmed to engage in certain behaviours, which may or may not be based on reason.

The question is not what choices will individuals make, but rather what genetic or social programming will survive in the long term.

John Maynard Smith

What behavioural patterns will be weeded out by evolutionary forces?

George Price

Hawk-Dove Game

One useful tool with which to examine issues of social and genetic programming is the Hawk-Dove Game. This is widely used in evolutionary biology as a starting point for thinking about behavioural patterns of animals and was introduced to the field by John Maynard Smith and George Price. The game highlights the importance of **evolutionary stability**, which examines which types of behavioural patterns are likely to survive evolutionary forces.

For simplicity, the game assumes that there are two types of animals in a species: "**hawk**" and "**dove**". The hawk-type fights if necessary when competing for a *prize*, such as a *mating opportunity* or a *scarce resource*. The dove-type makes an aggressive display, but falls short of non-ceremonial physical conflict.

DOVE

HAWK

It is useful to assign arbitrary payoff numbers for each potential outcome. In evolutionary biology these payoffs serve to show us each type of animal's **evolutionary fitness**. Access to the contested prize improves the animal's *prospects of reproduction or survival* (whether the prize is a mating opportunity or scarce resources). The higher the payoff, the better evolutionary fitness the animal has.

If a hawk-type animal comes into conflict with a dove-type animal, the dove-type backs down and receives a payoff of zero, while the hawk-type gets a payoff of 20, which is the value of the prize.

If both animals are dove-type, they have an equal chance at the prize. So, each gets the prize with a 50% chance, resulting in an expected payoff of (20/2) = 10.

If both animals are hawk-type, there is a physical conflict between them. Each animal has a 1/2 chance of winning the prize, which has the value 20. The animal who loses the fight gets injured and suffers an evolutionary fitness cost of -C. So, each animal has an expected payoff of:

$$20/2 - C/2$$
$$\rightarrow (20 - C)/2$$

These potential outcomes can be written in a payoff matrix.

WOLF B

	Hawk-type	Dove-type
Hawk-type	A:(20 - C)/2 B:(20 - C)/2	A:20, B:0
Dove-type	A:0, B:20	A:10, B:10

(Wolf A labels the rows)

When I meet another hawk-type we fight and I am likely to get hurt!

The Hawk-Dove Game with small cost of conflict

Examine the Hawk-Dove Game when the cost of conflict is less than the value of the prize – suppose that the cost of conflict (C) is 8.

Wolf B

		Hawk-type	Dove-type
Wolf A	Hawk-type	A:6, B:6	A:20, B:0
	Dove-type	A:0, B:20	A:10, B:10

If the animals were to choose their behaviour rationally, then hawk-type behaviour would be the dominant strategy – whatever the other animal is doing, it is always better to adopt a hawk-type behaviour.

> If animals were rational and were able to choose their type, the single Nash equilibrium would be for both animals to adopt a hawk-type behaviour, resulting in excessive conflict. The spirit of the game would be the same as the Prisoner's Dilemma.

Let's come back to a key tenet of evolutionary game theory and suppose that animals are **not making rational choices**, but are simply following their genetic or social conditioning.

John Maynard Smith

Suppose that there is a large population of animals, some of which are genetically or socially conditioned to engage in hawk behaviour and some of which are conditioned to engage in dove behaviour. The individual animals in this population are then randomly matched with each other to play the game.

An animal conditioned as a dove gets a payoff of zero if it is matched with a hawk-type or 10 if it is matched with a fellow dove-type.

An animal conditioned as a hawk gets 6 if matched with a hawk-type or 20 if matched with a dove-type.

When the cost of conflict is less than the value of the prize, animals that behave aggressively do better than less aggressive animals, whichever type they get matched with.

The Hawk-Dove Game can yield insights into the evolution of species. Access to a contested mate or food increases an animal's prospects of reproduction or survival and the loss from conflict decreases these prospects. Animals with higher evolutionary fitness (higher payoffs) are more likely to survive and reproduce.

If there is small cost of conflict, hawk-type aggressive animals do better than more peaceful dove-type animals of the same species. So **survival of the fittest** predicts that the entire species will eventually consist entirely of hawk-type animals.

Charles Darwin

Evolutionary forces drive out all dove behaviour. With all hawks, there is excessive conflict. Each member of the species gets a payoff of only 6. If every animal were conditioned to engage in dove behaviour instead, they would each get a payoff of 10. So, hawk behaviour is not optimal for a species as a whole.

Evolutionary forces do not necessarily lead to the best outcome for a species. Competition for scarce resources often means that individual benefits and group benefits are in opposition. Whenever this is the case, the species will evolve to maximize individual benefits at the expense of group benefits.

The tension between group benefits and individual benefits is present in physical traits as well as in behavioural patterns. Similar evolutionary forces can also affect how physical traits evolve. One example of this is **Cope's Rule**, named after American palaeontologist **Edward Drinker Cope** (1840–97), which says that a species generally increases in size over time.

If large male elephants are more likely to reproduce than small males, elephants will become larger over time. They can even become excessively large, decreasing the species' fitness. Scientists have also found that marine animals have been generally increasing in size over the last 500 million years, although the exact causes are still controversial.

One brake on this evolutionary process is the rise of a rival species which contests the same ecological resources: if a species becomes too inefficient due to its large size, it will eventually be pushed out by a more efficient competing species.

But then the cycle may repeat, with this new species also facing a conflict between individual benefits and group benefits, potentially becoming less efficient over time.

The Hawk-Dove Game with large cost of conflict

The evolutionary process is even more interesting in cases where the cost of conflict (C) is very high relative to the value of the contested prize. Suppose C = 24, and the value of the prize remains 20.

Lion B

		Hawk-type	Dove-type
Lion A	Hawk-type	A:-2 B:-2	A:20, B:0
	Dove-type	A:0, B:20	A:10, B:10

The high cost from losing a physical fight significantly changes the evolutionary fitness prospects of the hawk-type.

Suppose that initially some fraction "*p*" of the population is conditioned to behave very aggressively, adopting hawk behaviour. The remaining fraction "(1 - *p*)" of the population is conditioned to adopt dove behaviour. The fraction "*p*" can be as low as zero (none of the population are hawks) and as high as one (all of the population are hawks).

Since a dove-type never engages in a costly fight for resources, its situation is the same as it was in the case with a low cost of conflict. But it will be useful to examine its expected evolutionary fitness in more detail.

A dove comes in conflict with a random member of the population. With probability "p" its rival is a hawk. In this case, the rival gets the prize and the dove gets a payoff of zero.

But with probability ($1 - p$) the rival is also a dove. In this case, there is no physical conflict and the animals have an equal chance of getting the prize. The dove gets a payoff of 10.

So, the expected evolutionary fitness of a dove-type is the sum of the chance of meeting each type of rival times the payoff if that meeting takes place:

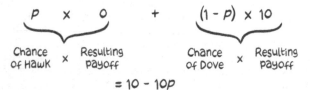

$$p \times 0 + (1 - p) \times 10$$

| Chance of Hawk | × | Resulting payoff | | Chance of Dove | × | Resulting payoff |

$$= 10 - 10p$$

Consider the situation of a hawk-type: this animal fights aggressively, even to the point of risking serious injury to itself.

With probability "p" the hawk is matched with another hawk-type and they fight. The cost of conflict is so high that it outweighs the benefit of capturing the prize and they both get an expected payoff of -2.

With probability $(1 - p)$ our hawk is matched with a dove-type. This rival backs down in the face of aggressive behaviour, so our hawk captures the entire prize without physical confrontation and gets an evolutionary fitness payoff of 20.

So the expected fitness of a hawk is the sum of the chance of meeting each type of rival times the payoff if it meets that type:

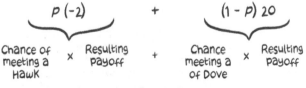

$$p (-2) + (1 - p) 20$$

| Chance of meeting a Hawk | × | Resulting payoff | + | Chance meeting a of Dove | × | Resulting payoff |

$$= 20 - 22p$$

If the evolutionary fitness of hawk-type lions is greater than the evolutionary fitness of dove-type lions, then on average hawks will survive and reproduce at a greater rate than doves. Hence, over time the fraction of hawks in the population will increase.

Using the calculations from the last two pages, a hawk's expected evolutionary fitness is greater than a dove's expected fitness, when:

$$20 - 22p > 10 - 10p$$

This can be rewritten as:

$$10 > 12p$$
→ $$10/12 > p$$
→ $$5/6 > p$$

If the fraction of hawks (p) in the population is less than 5/6, then the chance of a hawk running into another hawk and fighting is small enough that it is outweighed by the benefit from getting all of the prize when matched with a dove. So, over time the fraction of hawks (p) will increase due to evolutionary forces.

If the proportion of hawk-types in the population is greater than 5/6 (that is, $p > 5/6$), then doves will survive and reproduce at a greater rate than hawks, and the fraction of hawks in the population (p) will fall.

> If there are enough doves in the population, then my high-risk, aggressive nature pays off. But if there are too many other hawks, I get into so many fights that I can't keep my evolutionary fitness up.

In the long run, evolutionary forces will cause the proportion of hawk-types in the population to tend toward 5/6 and the proportion of dove-types to tend toward 1/6. These exact proportions are due to the particular numbers we used in the payoff matrix. But whenever the cost of conflict is higher than the value of the prize, evolutionary forces will drive the population to a point where both hawks and doves coexist.

In the long run, hawks and doves will coexist in the population at a ratio of 5 to 1 and both will do equally well on average. Hawks will appropriate all the resources when matched with doves, but they will have a high likelihood of getting seriously injured when matched with fellow hawks. Doves will lose the resource when matched with hawks, but will never be injured.

This long-run evolutionary "steady state" with the proportion of hawks in the population equal to 5/6 is called an **evolutionarily stable equilibrium**. It is an equilibrium which is stable in the sense that if we add a small number of animals with different conditioning, evolutionary forces will eventually restore the equilibrium.

HAWKS

DOVES

In general, evolutionary games are rich with possible outcomes. In our Hawk–Dove Game, there is a single evolutionarily stable equilibrium and the long-run steady state will eventually be restored regardless of how many animals with different conditioning we add.

But some games have more than one evolutionarily stable equilibrium. In these games, evolutionary forces will restore the equilibrium proportions if there are small changes to the population. But large changes in the population composition can cause evolutionary forces to bring the population to another equilibrium altogether.

If another large herd joins us, we could end up in a different evolutionarily stable equilibrium. In future generations we might evolve very different characteristics.

Some games have no evolutionarily stable equilibrium. In these games, the population will never settle down to a stable steady state. Rather, they will go through cycles, with the fractions of different types of animals endlessly rising and falling.

Evolutionary stability as an equilibrium refinement

Oddly enough, the evolutionarily stable proportion of hawk-types (5/6) is also equal to the equilibrium probability in the mixed-strategy Nash equilibrium of the game if the animals were choosing their strategies rationally. This is not a coincidence. To calculate the equilibrium probabilities in the mixed-strategy Nash equilibrium, we look for probabilities where players are just indifferent between the hawk and dove strategies. In equilibrium their expected values from both strategies are equal.

In the Hawk-Dove Game, we have the same level of expected evolutionary fitness of both types of animals at the evolutionarily stable equilibrium ratios. If their fitness differed, evolutionary forces would have one type prosper and the other languish until a stable state is reached.

Mathematically the two problems, the problem of rational decision makers...

...and the problem of genetically conditioned animals subject to evolutionary forces, are identical.

In the Hawk–Dove Game, the evolutionarily stable equilibrium gives the proportion of hawk-type and dove-type animals in the population. This is similar to the interpretation of the mixed-strategy equilibrium in the Tax Evasion Game. There the equilibrium gives the proportion of tax evaders in the population when players make rational choices.

In an evolutionary environment, focusing on the evolutionarily stable equilibria is a reasonable way to rule out equilibria which could not survive even small changes to the underlying population.

Sequential-move games

Often players can observe the actions of others before making their own moves. In some games, there is an order to players' actions. These are called **sequential-move games**. Most board games, such as chess, have alternating sequential moves.

For example, an entrepreneur thinking about whether or not to open a coffee shop on a particular corner can observe which other shops are already there and will consider which others may come to the same corner in the future if she were to start a business there.

Sequential-move games are **dynamic**, in the sense that the players can make decisions based on their observations of past actions and on their anticipation of future actions. Players conjecture what the other players would do in response to their possible choices, and then work backwards from the end of the game in order to decide what to do.

I knew having two shops here would scare away the competition. That's why I opened them last year.

A dynamic Battle of Sexes Game

We can examine the issues that arise in sequential-move games by creating a dynamic version of a simultaneous-move game. The Battle of the Sexes Game is a useful example.

In the standard Battle of the Sexes Game, Bob and Amy independently and simultaneously decide where to go for the evening. They want to be together, but they each have a different preferred activity. Remember the strategic form of the original, simultaneous-move Battle of Sexes Game?

Bob

	Football	Dance class
Football	A: 5, B: 10	A: 0, B: 0
Dance class	A: 0, B:0	A: 10, B: 5

Amy

Now, let's change the story slightly. Suppose that Amy gets out of work an hour before Bob. She goes to the location of one of the events and calls Bob telling him where she is. Once she does this, it is too late for her to change her location, but Bob can still make it to either one of the locations.

I am the first mover. I get to decide before Bob.

The extensive form of the game

Amy is the **first mover** and Bob moves second, having observed Amy's choice. The strategic form representation of the game is no longer as helpful as it was when the players made simultaneous moves, as the strategic form doesn't capture the order of the choices. For this we need a new diagram to represent the sequential-move game: the **extensive-form** representation. This is also called the **game tree**.

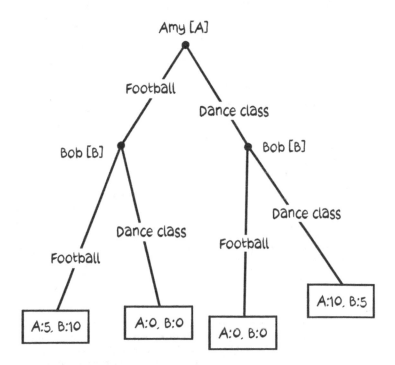

The extensive form introduces the order of choices through the use of **decision nodes**, dots that represent a point at which a decision can be made.

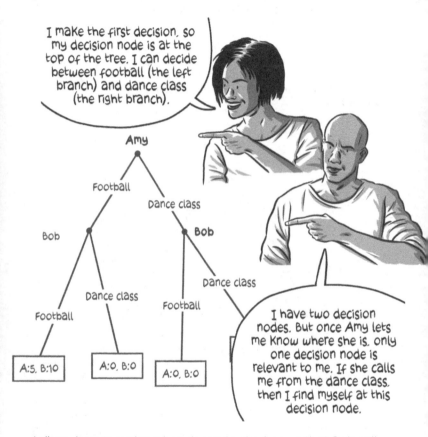

When Amy is making her decision she knows that Bob will be able to observe her choice before making his own. She also knows that her decision will influence his choice. So, she will try to figure out how he would respond to each of her possible choices.

Subgame perfection

If Amy were to call Bob from the football stadium, then for Bob only the lower left-hand decision node would be relevant. So we can think of the game from that point on as a game in itself. This is known as a **subgame**. At that point, Bob would simply do the best he could from then on.

If Amy calls from the football match, I can either choose football and get a payoff of 10 or choose dance class and get zero. I'm going to football.

If I call Bob from the football stadium, then Bob will come to football; so I'll get a payoff of 5 if I choose football.

Bob

Dance class

Football

A:5, B:10

A:0, B:0

Amy will also consider what Bob would do if she decided to go to the dance class. If she were to call Bob from the dance studio, Bob would be faced with a different subgame (the right-hand side of the extensive form).

Amy and Bob's dynamic game is solved by backward induction. Amy conjectures what will happen at the end of the game and works backwards from there to figure out her best choice.

It is rational for Amy to choose dancing, since she knows that Bob will follow her to dance class. This is a **subgame-perfect Nash equilibrium**: players have best responses to each other in each subgame of the original game. Subgame perfection implies that players are forward-looking. They do the best they can at each decision node they encounter without either holding grudges or developing goodwill for past actions.

In this game, the subgame-perfect Nash equilibrium is particularly beneficial for Amy. Here there is an advantage to being the first mover.

> I know that Bob will follow me anywhere, so I might as well go to the activity I like best. It's perfect.

> Well, it's subgame-perfect, at least.

This game has **first-mover advantage**, but not all sequential-move games have this feature. There are many games where making an early commitment puts a player at a disadvantage.

Non-credible threats

Most people find the subgame-perfect Nash equilibrium where both players go to dance class the most plausible Nash equilibrium, but it is not the only one.

For instance, Bob could declare that he will always go to football regardless of what Amy chooses. If Amy believes this, then she will expect to end up alone if she goes dancing. So, she would choose football, since she always prefers being with Bob to being alone. This is also a Nash equilibrium, but it relies on Amy believing that Bob would follow through with his threat and would go to football even when Amy calls him from the dance class. This wouldn't be in his own best interest; therefore Bob's threat is not credible.

Subgame perfection discards Nash equilibria that depend on players making *non-credible threats* and *non-credible promises*.

Credit markets

The interaction between lenders and borrowers can be modelled as a sequential-move game. This can be useful for analysing why some good projects fail to get financed.

The extensive form of the game gives the expected payoff numbers (profit in millions of pounds) to a loan applicant (A) and a bank (B). For simplicity, assume that the bank and the loan applicant have full information about the game tree and the expected payoffs from each project.

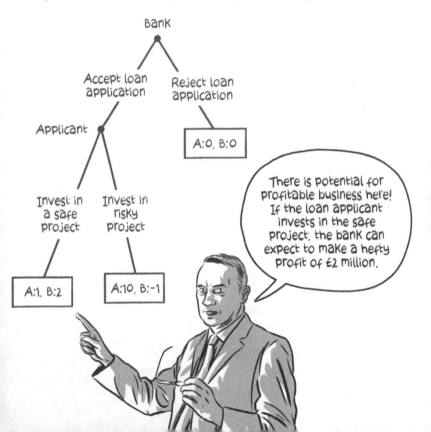

The applicant has the choice of investing in either a safe project or a risky project, but only if the bank gives credit.

The bank would like the loan applicant to invest in the safe project. However, it cannot *monitor* the applicant's daily business decisions; therefore it cannot dictate which project the applicant invests in. In the subgame-perfect Nash equilibrium the bank manager refuses the loan application, even though both the bank and the loan applicant could have benefited from a profitable business deal with the safe project.

> I reject your loan application because I know that you would invest in the risky project if I were to give you the credit. In that case my expected payoff would be negative. Simple backward induction!

> But then we both get zero payoff!

The applicant may promise the bank that she will invest the funds in the safe project. And she may be sincere – after all, she gets zero payoff if the loan application is rejected but a payoff of £1 million from the safe project if the application is accepted.

However, if the bank were to extend the credit, once the applicant received the funds, she would compare her expected payoff from the safe project to her expected payoff from the risky project. So, she would choose the risky project, breaking her promise. This is called the **time-inconsistency problem**: the decision maker does not find it optimal to follow the original action plan.

Once you have the money in hand you won't be able to resist the potential gains from the risky project. And I certainly do not want to fund a risky business.

I wish I could make a credible promise to stick to the safe project.

The subgame-perfect Nash equilibrium where the bank rejects the loan application is not Pareto efficient. Both the applicant and the bank would earn higher expected payoffs if the safe project were financed.

What if the applicant could find a way to *credibly commit* to the safe project, such that even if she could go for the risky project, she would choose not to?

Financial markets often use *collateral* as a **commitment device**. For instance, the applicant could use her family home as collateral. As long as losing the family home would be costly enough for the applicant (either financially and/or in psychological suffering), the collateral changes the expected payoff of the risky project to the applicant. So, she would choose to invest in the safe project. Hence the bank would approve the loan.

The house is worthless to the bank by the time you deduct the legal fees to liquidate it if she fails in her business. Why do you give her the loan?

The house is worthless to the bank, but it is very valuable to her. It's her family home; she won't risk losing it. She will go with the safe project.

Microcredit

If loan applicants are able to provide collateral to credibly commit to a safe project, they gain access to credit markets to finance their businesses. However, those who do not have existing assets to use as collateral will have their loan applications rejected in the subgame-perfect Nash equilibrium because of the *time-inconsistency problem*.

Due to the difficulty of coming up with a commitment device, the poor stay poor, while the rich get richer. Lack of access to credit markets can prevent poor people from being upwardly mobile, which can cause severe social unrest and violence.

Bangladeshi economist **Muhammad Yunus** (b.1940) was awarded the 2006 Nobel Peace Prize for his answer to this problem – founding the Grameen Bank and pioneering the concept of **microcredit** to help the poor gain access to financial markets.

'You cannot create the bank of the poor with the same architecture as the bank of the rich.'

To allow access to credit markets, Yunus solves the problem of lack of a commitment device by giving the poor microcredit (small loans) in a *pooled* manner – loans to a connected group of people rather than to an individual. Each applicant in the pool then makes sure that the other applicants invest in safe projects.

This is a poor villager with no collateral. If you give her credit, how do you know that she won't run to the casino tonight and lose it all?

Her loan is tied to the loans I am giving to two of her neighbours from the same village. If she cannot pay us back, the other two know that I will never lend to them again. So, they will make sure that she puts the money to good use.

Nuclear deterrence

Since WWII the two major nuclear powers, the USA and Russia, have both adopted a policy of nuclear deterrence based on mutually assured destruction. The idea is that if one side attacks, the other side can retaliate with overwhelming force, destroying the aggressor. Hence, no one would attack in the first place.

To date there has not been a global nuclear war, so this policy has been successful. However critics argue that the desired equilibrium may not be subgame perfect. It may be based on non-credible threats. If this is the case, there may be trouble ahead.

The mutually assured destruction strategy is based on the idea that if enemy missiles were inbound, policy makers in the targeted country would retaliate, destroying the aggressor. However, retaliation does not alter the situation of the targeted country: their fate is sealed by the incoming missiles.

The leader of the attacked country will almost certainly want revenge. In that case, retaliation will be the best choice in the subgame when they are deciding whether or not to counterattack. If so, and the enemy knows it, then the desired equilibrium where there is no attack in the first place is subgame perfect. There will be no nuclear war.

However, the person deciding whether or not to retaliate may have moral reservations about killing millions of civilians. After all, once the enemy missiles are in the air, there is nothing to gain by retaliating. Therefore, in the retaliation subgame, a morally concerned decision maker will choose not to counterattack.

In that case, the threat to retaliate is not credible. The desired equilibrium where there is no initial attack is not subgame perfect. There is no reason not to launch a pre-emptive strike, since the attacked party will not retaliate.

If a policy maker is concerned about the moral implications of killing millions of innocent people, how can they avoid becoming a target of a nuclear attack?

One potential solution to this problem is to **delegate** the retaliation decision to someone who is likely to be motivated by revenge or by a duty to follow the prearranged procedure. This will ensure that retaliation is a credible threat.

A second way of making the threat of retaliation credible is to give many individuals the option of initiating an overwhelming strike. This is the **proliferation** solution. Then, when the enemy is contemplating an attack they must gauge the chance that at least one of these people is motivated by revenge. The more people who have the ability to launch a retaliatory strike, the more likely it is to happen. If retaliation is likely, then there will be no attack in the first place.

Each of you has the code to launch a counterattack!

In practice, both delegation and proliferation are used to make retaliation credible and, hence, to ensure that no initial attack is part of a subgame-perfect equilibrium.

Hollywood has suggested a third option to solve this problem: the retaliation decision could be completely automated, guaranteeing a counterattack. This is the premise of the "Doomsday Device" in *Dr. Strangelove*, the "War Operation Plan Response" in *WarGames* and "Skynet" in the *Terminator* films.

The extent to which such an approach has been taken in reality is unclear. However, as pointed out in *Dr. Strangelove*, a Doomsday Device is only useful as a deterrent if the potential attacker knows of its existence. As soon as it needs to be used, it has already failed its purpose. Hence, there is no reason to keep the existence of such a device secret and every reason to publicize it. Therefore, we can be pretty sure that this is not yet an approach the major superpowers have taken.

Information problems

In the extensive-form games discussed so far, players have full knowledge of the game tree. However, situations often arise where players have **incomplete information**: they might not know all of the other player's available strategies or potential payoffs. Players may not be sure what kind of person they are dealing with or what their motivations are.

I am trying to decide if we should accept her offer to go into business together, but I cannot tell if she is trustworthy or not. What if I accept her offer and it turns out that she is a cheat?!

There are also situations where players have **imperfect information** about the game tree: players' past actions may be either unobservable or imperfectly observable. This means that players do not know exactly which decision node they are on in the game tree.

In all walks of life people make decisions where their information is either imperfect or incomplete, or both. This has important implications for the strategic interaction between players, especially if one side is better informed than the other.

Asymmetric information

American economists **George Akerlof** (b.1940), **Michael Spence** (b.1943) and **Joseph Stiglitz** (b.1943) were awarded the 2001 Nobel Prize for their analyses of markets with **asymmetric information**: where a player has superior information compared to others.

For instance, in the car insurance market a driver has *private information* about their own driving habits. The insurance company has *incomplete information*: it doesn't know the driver's habits, so it doesn't know the payoffs it will get from selling them insurance.

A manager may have *imperfect information* about an employee's habits. If the employee doesn't make progress in a task, the manager doesn't know whether to blame the employee or to believe that it's a particularly difficult task.

I'd give you up to £6,000 if I were certain that this car was reliable, but I'm not. Sorry, £2,000 is my last offer!

I would have accepted £6,000, and we both would have been happy because my car is indeed very reliable. As I won't be able to get a good price, I better take the car off the market and keep it as a back-up.

Asymmetric information and unemployment

Macroeconomists, who are interested in large-scale economic patterns and effects, often study the problem of persistent **unemployment**, a situation in which there are people who are willing to work but who cannot find a job.

Persistent unemployment presents a puzzle for standard economic analysis: if there is unemployment, then for each job opening there are many applicants. In that case, firms can offer lower wages and still fill all vacancies. With lower wages, hiring is cheaper and firms employ more workers. It seems like wages should eventually adjust downward to the point where the number of people who want to work equals the number jobs.

Given this expected pattern, why is there persistent unemployment? Why don't wages simply adjust downward until there is no more unemployment?

Nobel Prize winner Joseph Stiglitz and American economist **Carl Shapiro** (b.1955) show that one of the causes of persistent unemployment in the economy is **hidden action** at the workplace –where the actions of workers are not perfectly observable.

Consider a worker on a fixed wage. He can either work hard or he can be lazy and shirk his responsibilities. The worker's effort is not perfectly observable. The manager would fire the worker if he were caught shirking, but she can only monitor him imperfectly, so she will not always catch a shirking worker.

When the worker decides whether or not to shirk, he compares the benefit to the cost of shirking. The benefit is a more enjoyable work day. The cost is the chance of being caught combined with the value of what the worker loses if he is fired.

If there is no unemployment and the firm offers the same market-clearing wage as other firms, then the worker will try his chances with shirking.

In order to encourage workers to work hard, a firm has to give them something to lose if they get caught shirking. It can do this by offering higher wages than workers could get elsewhere. These high **efficiency wages** can induce efficient output from the workforce.

"To induce its workers not to shirk, the firm attempts to pay more than the "going wage"."

Joseph Stiglitz

Efficiency wages pose a problem, however, in that all firms are faced with the same incentive to offer higher wages to induce efficiency. But if each firm introduces a higher wage, wages increase in the marketplace. Since more people will be willing to work at those higher wages and the number of jobs has not gone up, this will result in unemployment.

In this case, employees will be motivated to work hard because if they lose their job it might take them a long time to find a new one.

More on asymmetric information

Situations often arise in which people are not sure about what kind of person they are dealing with. These are games of *incomplete information*, where players are uncertain about the characteristics of the other player and so are uncertain about the payoffs from the possible outcomes of the game.

Often this is represented by thinking of the other player as having a *type*. Each type is associated with different payoffs from the possible outcomes of the game. A player typically knows his or her own type, but it is unknown to the other player. Hence there is *asymmetric information*.

We have a great new technology to make a car that won't need any technical assistance for 20 years. I suggest we introduce it to the market next year.

Consumers won't know our improved car is as reliable as it is. So we won't be able to charge a high enough price to cover the cost of the new technology. Shelve the production plans for now.

Signalling product quality

It is difficult for an individual consumer to discern the quality of a product before purchasing it. However, the company has a good idea whether its product is likely to be long-lasting or not. The seller knows their type (either high-quality or low-quality), but the buyer cannot tell.

What consumers need is a way to infer which firms are offering high-quality products, and which are not. Of course, a firm has every incentive to claim that its product is of high quality, whether it is or not. So, direct statements by the firm have no value.

For some products, firms can overcome the asymmetric information problem by offering free samples. Yet for other products this may not be possible.

If it can't provide believable, direct information about product quality, a high-quality firm may need to find a device to **signal** its quality to the consumer. For this to work, the signal must be some observable action that a high-quality firm can take but that would not be worthwhile for a low-quality firm.

Warranties as a signalling device

Even if a consumer does not plan on keeping their proof of purchase, the existence of the warranty can convince the consumer to purchase the product because it signals quality.

The consumer can reason that only a company with a reliable product could afford to offer a long-term warranty, since there would be relatively few claims made.

A firm with an unreliable, low-quality product would realize that there would be many warranty claims, and so a long-term warranty would be too expensive for them to offer.

In a **separating equilibrium**, a firm's choice of warranty depends on the type of its product (high-quality or low-quality). A firm with a high-quality product chooses to offer warranty, but a firm with low-quality product chooses not to. Through *self-selection*, the warranty allows a high-quality product firm to differentiate itself from a low-quality product firm. The two different types *separate* themselves through their behaviour.

Advertising as a signalling device

Advertising may also serve as a quality signal if the firm is selling a repeat-purchase product, such as shampoo. This is because the return on investment from advertising is different for products of different quality.

While consumers may not know the quality of the product before the first purchase, after having used it once, they can assess its quality. If the firm is selling a low-quality good, new customers who have seen an advert for the product will buy it only once. They will then realize that it is of low quality and won't buy it again. However, if the firm is selling a high-quality product, new customers will become repeat purchasers.

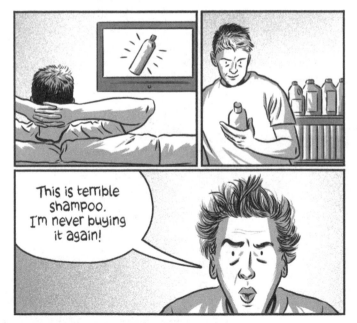

Advertising works as a signalling device because of a simple cost–benefit analysis. The cost of advertising is the same no matter what quality of product the firm sells. But the benefit from advertising is much higher for a firm producing a high-quality product since new customers will make repeat purchases. For the same advertising expense, a firm selling a low-quality product can get only one-off purchases from first-time customers. Therefore, a high level of advertising can only be profitable for a firm with high-quality goods.

Consumers observing an expensive advertising campaign can infer that the firm would only have advertised if it knew that its product would generate repeat purchases. Hence consumers use advertising as a signal for high quality.

Religious ritual as a signalling device

British-Israeli economist **Gilat Levy** (b.1970) and Israeli economist **Ronny Razin** (b.1969) show that religious ritual can serve as a signal for a practitioner's genuine religious belief. Many religions encourage a connection between spiritual belief and social behaviour. Members of religious communities are often found to behave more cooperatively with each other than they behave with non-believers. This provides a material, as well as a spiritual, benefit to being a member of such a community.

But this works only if the members of the community know that they have shared beliefs as a basis for their interaction. Since membership of the community provides some material benefits, non-believers have an incentive to pretend to be true believers.

To avoid fake signals from non-believers, religious groups develop rituals that are difficult to adhere to – such as distinctive clothing, public prayers and dietary restrictions. As true believers receive both material and spiritual benefit from group membership, for them it is worth doing difficult rituals.

Ronny Razin

Non-believers get only the material benefit of group membership. So if the ritual is too difficult it will not be worth doing for them. Ritual can signal to community members that a person is a true believer.

Gilat Levy

Decision making in groups

So far, we have examined situations in which each player makes a decision for him- or herself.

However, often decisions are made by a group of players. While an individual player may have input into the group decision, not all members of the group might agree on the best course of action. When everybody cannot get their first choice, it may be difficult to agree on what the group's preference is.

I'm not really sure the best way to reconcile everyone's preferences in the group...

Studying group behaviour presents a challenge for game theory because the group as a whole may seem *irrational* even when each member of the group is *rational*.

Rational decision makers have **transitive preferences**. This means that if a decision maker prefers alternative A over alternative B and prefers alternative B over alternative C, it must be that he or she prefers A over C (the symbol ">" means "preferred to"):

So A > B and B > C implies A > C

However, even when all group members are rational, group preferences can be **non-transitive**.

That is, for groups, A > B and B > C does not necessarily mean that A > C!

You are a man of logic. So are we all.

So, how come when we get together, our group preferences don't seem to make sense?

We can see non-transitive group preferences in action in the example of a city which owns a vacant lot. There are three proposals for what to do with the land. It could be used as a park, as a recycling centre or as a new school.

The city council has to decide which of these options to choose. There are three people on the council. Each council member individually prefers a different alternative as their top choice.

	Mr Peters	Ms Reynolds	Mr Singh
First choice	Park	Recycling	School
Second choice	Recycling	School	Park
Third choice	School	Park	Recycling

Children are our future! We need to build a school.

We must ensure our planet's future – which is why we urgently need to build a recycling centre.

Let's vote to see which option has the most support.

In a series of votes, the committee compares two options at a time. Suppose that each council member votes for the option that he or she actually prefers, which is referred to as **sincere voting**.

The committee votes 2–1 for a school over a park. As a group they prefer school over park, so they definitely should not build a park. For the group:

School > Park

The only thing then left to decide is whether they should build a school or a recycling facility.

The committee votes 2–1 for building a recycling centre over building a school. For the group:

Recycling Centre > School

The matter is settled; the committee thinks that a school is better than a park and a recycling centre is better than a school.

In the vote between recycling centre and park, Mr Peters votes for park (his first choice), Ms Reynolds votes for recycling (her first choice), and Mr Singh votes for park, since recycling is his least favourite option. The committee thinks that a park is better than a recycling centre by 2–1.

Park > Recycling centre

Every person on the committee has transitive preferences and votes sincerely. But when acting as a group the committee's preferences are **non-transitive** – whatever it chooses, the group will always think that another option is better.

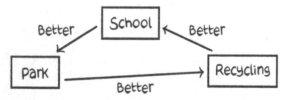

American economist **Kenneth Arrow** (b.1921) was awarded the Nobel Prize in Economics in 1972 for a mathematical result known as "Arrow's Impossibility Theorem". It shows that, for groups which are not run by a dictator, there will always be the possibility of some situations where group preferences are non-transitive, where we reject a choice that may be better for everybody or where irrelevant options change our choice. These problems are inherent in group decision making.

´Attempts to form social judgements by aggregating individual expressed preferences always lead to the possibility of paradox.´

Arrow's Impossibility Theorem makes a lot of the strange behaviour we see in committee meetings and in parliaments more understandable. For instance, in committee work we frequently see the same issue coming up over and over again.

There are many ways to arrange group decision making. These range from autocracy, where one person makes all the decisions based on his or her own preferences, to classical democracy, where all group members have an equal say in the decision – with an infinite variety of systems in between these extremes.

Arrow's Impossibility Theorem shows that, other than autocracy, whatever system we use to decide our best choice as a group, there is always the possibility that the group behaves inconsistently.

Where we've come from ...

While game theory took hold as a field of study in the 1940s, its central themes of cooperation and conflict are as old as human society.

For example, in his book *Leviathan*, English philosopher **Thomas Hobbes** (1588–1679) argues that:

> In the absence of a strong government, life would be "nasty, brutish, and short".

His argument is essentially game theoretic in nature: without a strong government to enforce contracts, cooperation would break down because each person would be worried that the other was immoral. This would also lead to violence.

> If he reneges on his promise, he will expect revenge. In that case he will kill me before I take revenge. Maybe I should kill him now.

> If he doubts I will keep my word, he might seek to harm me now. I ought to be ready to defend myself.

Examples of game theoretic reasoning can also be found as early as the writings of Plato, who reports a recollection of Socrates from the battle of Delium in 424 BC.

...and where to go from here

The development of game theory as a discipline has provided an extensive toolkit which allows us to explore conflict and cooperation in much greater depth.

We can now answer questions which were difficult, if not impossible, to address before, such as:

In the Hawk-Dove Game (p. 101-15), if global warming makes the resources a species is competing over scarcer, will there be more or fewer aggressive animals over time?

In the Currency Speculation Game (p. 71-3), does having a higher exchange rate increase or decrease the chance of a speculative attack?

In the Tax Evasion Game (p. 87-9), if tax rates go up, what will happen to your chance of being audited?

The mathematical background and presentation of much of game theory can make it difficult for newcomers to approach the field and acquire tools that would be useful to them. So, in this book we have deliberately avoided mathematical complication and focused on game theory's central ideas.

We have looked at examples where players have a limited number of choices. However, players often have to choose from continuous options. In these situations, the game theory logic is exactly the same, but the presentation becomes more mathematical.

For instance, we might look at a firm deciding on whether to advertise or not. In this simple representation, it chooses whichever option gives the higher payoff. The choice is binary: advertise or don't. In reality, the decision is typically *how much* to advertise. The firm's choice can be any level of advertising.

As you work with the tools you acquired from this book, you will eventually run into situations where deeper knowledge would be useful. Or you may be interested in learning additional game theoretic tools. A good source for this next step is:

The examples in Gibbons' book lean towards economics, but the tools are useful for any field.

*This same book is marketed in the US as *Game Theory for Applied Economists*, and is published by Princeton University Press.

Over the past 70 years, a wide variety of game theory tools have been developed for the analysis of strategic thinking. And indeed, many of these tools are quite technical in nature.

But you don't need the entire toolkit in order to do useful and interesting work with game theory. Just as you don't need every tool in the hardware store in order build a shelf, you do not need every tool in the game theory toolbox in order to gain useful insights into new situations with opportunities for cooperation or conflict. The tools you have already learned in this book are more than enough to provide useful insights.

About the Authors

Dr Ivan Pastine is a high-school and college dropout, whose game theoretic explanations of financial crises have been required reading in the PhD programmes at Harvard and LSE. He has been a handyman, a Boatswains Mate in the US Navy and, in more recent years, a lecturer at University College Dublin.

Dr Tuvana Pastine is a Turkish economist working at Maynooth University in Ireland. She specializes in applications of game theory and has published on a wide variety of fields, analyzing coordinating advertising and price dynamics, political campaign financing, affirmative action in education, sovereign default, labour migration and international trade.

Tom Humberstone is an award winning comic artist and illustrator based in Edinburgh. He contributed a weekly political cartoon to the *New Statesman* for three years and continues to produce comics and illustrations for *The Nib*, *Vox*, *The Guardian*, *Vice* and Image Comics among others. He listens to an absurd amount of podcasts.

Index

Index

Index